BETTER SEX

MAN ALIVE

BETTER SEX

Sarah Brewer, M.D.

MACMILLAN ◆ USA

Conceived, edited, and designed by
MARSHALL EDITIONS
170 Piccadilly
London W1V 9DD

Copyright © Marshall Editions Developments Limited 1997

First published in the U.S.A. in 1997
by Macmillan Publishing Company, a division of Macmillan, Inc.

MACMILLAN
A Simon & Schuster Macmillan Company
1633 Broadway
New York, NY 10019

Library of Congress Cataloging-in-Publication Data

Brewer, Sarah.
    Better sex / Sarah Brewer.
        p. cm. — (Man alive)
    Includes index.
    ISBN 0–02–861504–2
    1. Sex instruction for men. 2. Men—Sexual behavior. 3. Sex (Biology)
I. Title. II. Series: Man alive (New York. N.Y.)
HQ36.B686 1997
613.9'6'081—dc20   96–9700
                                                             CIP

10 9 8 7 6 5 4 3 2 1

Originated in Singapore by Classicscan
Printed and bound in Portugal by Printer Portugesa

Editor	Jonathan Hilton
Art editor	Vicky Holmes
Photographer	Laura Wickenden
DTP editors	Mary Pickles, Kate Waghorn
Copy editors	Jolika Feszt, Maggi McCormick
Indexer	Judy Batchelor
Managing editor	Lindsay McTeague
Production editor	Emma Dixon
Art director	Sean Keogh
Editorial director	Sophie Collins

CONTENTS

MORE SATISFYING SEX
Better sex is a goal any man can achieve with little effort.

Are you worried that the spark seems to have gone out of your sex life? Is your libido flagging? Is every sexual encounter boringly the same? Do you wonder why everybody else seems to have a terrific time in bed? Are you happy with your love life, but do you secretly wonder whether it could be even better?

If the answer to any of the questions above is yes, then this book has been written with you in mind. It will take you through several stages, each designed to help you enjoy a fuller, richer, more varied, and satisfying sex life. It includes information on:
♦ how to improve your sex drive
♦ how to prolong your lovemaking
♦ how to be a better lover
♦ how to delay orgasm
♦ how to boost the intensity and pleasure of your climax
♦ how to help your partner reach orgasm during penetration
♦ exercises to strengthen your vitally important pelvic floor muscles
♦ the positions that give the most satisfaction to men
♦ the positions that give the most pleasure to women
♦ the use and likely benefits of aphrodisiacs.

WHAT IS BETTER SEX?
Simple sex for procreation is a basic survival instinct – among humans as well as the rest of the animal world – but good, emotionally fulfilling sex is a learned skill that requires plenty of

practice. Learning to enjoy better sex has many facets.

To reach your full sexual potential, you may first have to get fit – if running for a bus leaves you gasping for breath, your performance in bed probably lacks a certain panache. Next, you have to get to know your own body, to understand your own needs, as well as the mechanics of lovemaking. And finally, you may need to fine-tune your understanding of your partner's needs.

Good lovemaking is not about hundreds of contorted positions – although it is useful to know which ones are good for deeper penetration, for better clitoral stimulation, or for when your erection is less than solid. Better sex is all about confidence, positive thought, and spontaneity. It means losing your inhibitions, learning to communicate with your partner, and being prepared to experiment a little.

In order to enjoy better sex, you ideally need to be in a loving, long-term relationship so that you are emotionally secure and involved in both giving and receiving pleasure. It is much easier to be experimental – and to have fun – with somebody you know well and love than it is with a relative stranger.

This book explores a number of ideas and techniques to help you improve your sex life. By following the tips and exercises, you will be well on your way to being a better lover and enjoying a more fulfilling sex life.

THE ROUTE TO BETTER SEX

Different men have different aims when it comes to improving their sex life. This chart directs you to specific pages to help you solve particular problems.

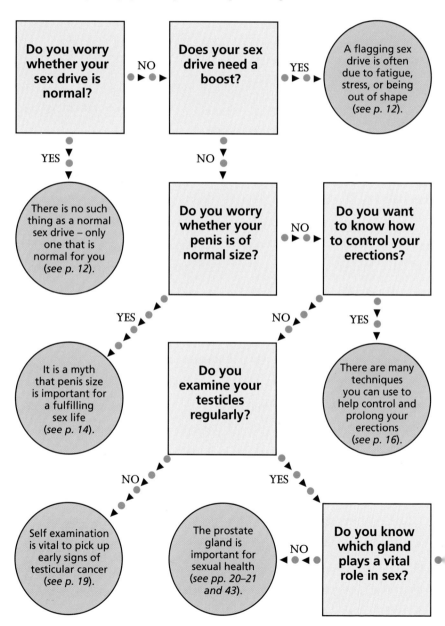

Do you worry whether your sex drive is normal?

NO ►

Does your sex drive need a boost?

YES ►

A flagging sex drive is often due to fatigue, stress, or being out of shape (*see p. 12*).

YES ▼

There is no such thing as a normal sex drive – only one that is normal for you (*see p. 12*).

NO ▼

Do you worry whether your penis is of normal size?

NO ►

Do you want to know how to control your erections?

YES ◄

It is a myth that penis size is important for a fulfilling sex life (*see p. 14*).

NO ◄

Do you examine your testicles regularly?

YES ▼

There are many techniques you can use to help control and prolong your erections (*see p. 16*).

NO ◄

Self examination is vital to pick up early signs of testicular cancer (*see p. 19*).

YES ►

The prostate gland is important for sexual health (*see pp. 20–21 and 43*).

NO ◄

Do you know which gland plays a vital role in sex?

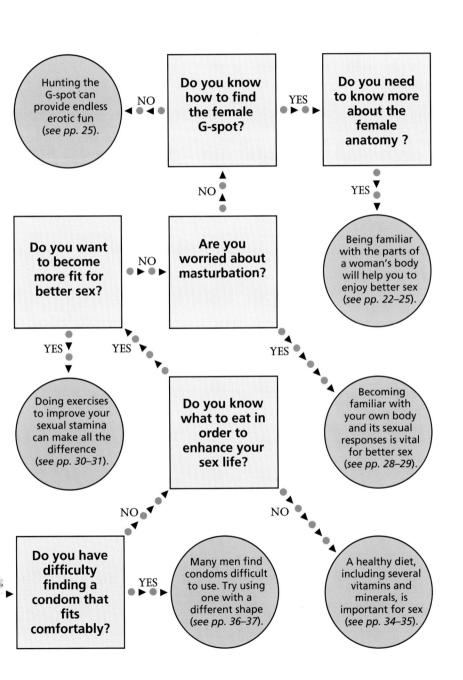

Hunting the G-spot can provide endless erotic fun (see pp. 25).

◄ NO ◄◄

Do you know how to find the female G-spot?

YES ►►►

Do you need to know more about the female anatomy ?

NO

Are you worried about masturbation?

YES

Being familiar with the parts of a woman's body will help you to enjoy better sex (see pp. 22–25).

Do you want to become more fit for better sex?

NO ►►

YES ▼

Doing exercises to improve your sexual stamina can make all the difference (see pp. 30–31).

YES ►

Do you know what to eat in order to enhance your sex life?

YES ►

Becoming familiar with your own body and its sexual responses is vital for better sex (see pp. 28–29).

NO

NO

Do you have difficulty finding a condom that fits comfortably?

YES ►

Many men find condoms difficult to use. Try using one with a different shape (see pp. 36–37).

A healthy diet, including several vitamins and minerals, is important for sex (see pp. 34–35).

THE PHYSIOLOGY OF SEX

This chapter sets out to give you an understanding of how your body works. By understanding the factors that affect your sex drive – and that of your partner – you can quickly restore your libido by taking action to deal with fatigue and stress.

The male reproductive system consists of the penis, testicles, prostate gland, and their interconnecting tubing. Many men take their sexual workings for granted, but there are simple dietary and lifestyle changes that can lead to better sexual health. Similarly, simple techniques can be used to help control your erection, postpone orgasm, and prolong your enjoyment of lovemaking.

Sexual arousal is triggered by a number of different stimuli, including smells, sounds, food, pictures, and physical contact, that provoke erotic thoughts. Despite massive research, this process of arousal is still not fully understood, but hormones are thought to play an important part.

By learning about the physiological changes that occur in your and your partner's bodies during foreplay and sexual arousal, you will be in a better position to judge when to move on to full penetrative sex, when and how to prolong and control the different pleasurable sensations, and how to increase your chance of enjoying a better, more satisfying sexual encounter.

WHAT IS A NORMAL SEX DRIVE?

This often-asked question has a simple answer.

Your sex drive is basically controlled by the male hormone testosterone, the amount of which varies from person to person. Everybody, therefore, has a different sex drive, and as long as both you and your partner are happy with the frequency with which you make love, you should consider your sex drive to be normal.

PRIMARY FACTORS

The frequency with which a man makes love depends on four main factors:

◆ The length of time he has been in a relationship – after five years, the frequency of sexual intercourse falls each year by an average of nearly one time per month.

◆ Overall health – a man in excellent health makes love on average nearly 11 times per month, compared with fewer than 9.5 times for those who feel that their health is only fair.

◆ The age of a man's partner – as she becomes older, he is likely to make love less often each month.

◆ Income – the more you earn, the less likely you are to have sex regularly, probably due to stress and general exhaustion.

LIFESTYLE FACTORS

Your sex drive can be affected by lifestyle factors, all of which are in your power to control. If you think your sex drive is fading, you should:

◆ Get plenty of sleep – tiredness will lower your libido.

◆ Avoid stress, since this lowers testosterone levels.

◆ Exercise more – this will boost metabolism, improve your fitness level, and relieve stress.

◆ Stop smoking – there may be a link between the number of cigarettes smoked and the rigidity of an erection.

◆ Cut down on alcohol – excessive drinking increases the rate at which testosterone is broken down.

◆ Take a multinutrient supplement – a lack of some vitamins and minerals can reduce sex drive.

◆ Check with your doctor to make sure that a low sex drive is not a side effect of prescribed medication.

If you think you have a low sex drive, it is important to seek help. As a first step, see your doctor, who may refer you to a sex therapist.

Male menopause

The highest levels of testosterone occur when men are in their teens and early 20s, and then gradually decrease after that. During middle age, the way in which men's bodies respond to testosterone may change so that they start to experience symptoms of the so-called male menopause. Symptoms may include fatigue, irritability, a lowered sex drive, aching joints, dry skin, difficulty sleeping, excessive sweating, hot flashes, and low mood. If you are age 45 years or older and have noticed these symptoms, consult your doctor.

THE PENIS

As the most potent symbol of a man's virility, the actual size of the male sex organ assumes a far greater importance than is warranted.

The size of a man's penis can vary considerably and still be considered within the "normal" range. In fact, as long as you are anatomically able to penetrate your female partner, it does not really matter what size or shape your penis is.

THE BASIC FACTS

Approximately nine out of ten men have penises that measure between 5½ in (14.5 cm) and 7 in (17.5 cm) when erect. The average size of the penis measured from tip to base is 6¼ in (16 cm). When flaccid, the size of the average penis ranges from 3 in (7.5 cm) to 6 in (15 cm), depending on the surrounding temperature. It generally lengthens by around 2 in (5 cm) when erect.

A penis that looks short when flaccid will often tend to lengthen proportionately more than longer ones. Although it is common for men to be sensitive about the size of their penis, it is very unusual to have one that is either abnormally long or abnormally short.

DOES LENGTH REALLY MATTER?

Size is less important to a woman than many men think. There is no relation between penis size and a man's ability to be a caring, considerate, and satisfying lover. Your qualities as a lover – your patience, generosity, and tenderness – contribute more to better sex than the size of your penis.

It is the stimulation of the clitoris, which lies outside the vagina, that triggers a woman's orgasm (*see p. 25*) – and this is possible whatever the size of her partner's penis. In fact, many women prefer a man with a slightly smaller penis because it is more likely to stimulate the clitoris while repeatedly entering and withdrawing from the vagina during lovemaking.

Similarly, since the vagina expands enough to allow for the passage of a baby during birth, no man has to worry that his penis is too big for his partner. If, however, you are large, you need to make sure that there is good lubrication to prevent soreness of both penis and vagina due to friction.

IS SHAPE SIGNIFICANT?

Research suggests that a penis that is thicker than average at the base may

IN AN UNCIRCUMCISED PENIS
the protective foreskin, which rolls back to reveal the glans, is intact.

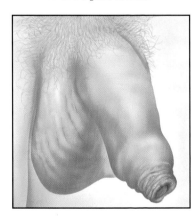

be able to stimulate the woman's clitoris more easily than a thinner penis can. A man with a thin penis need not worry, however. A study using an ultrasound probe to measure movement of the female clitoris during vaginal intercourse found that the greatest amount of clitoral movement occurred when a man entered his partner from behind, with the woman lying on her side (*see pp. 64–69 for different positions*). This was true whatever the size of the man's penis.

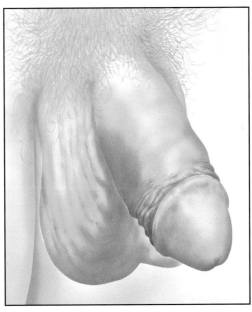

DURING CIRCUMCISION
the foreskin covering the tip of the penis is surgically removed, exposing the sensitive glans.

CIRCUMCISION

In addition to being a religious rite of Jews and Muslims, it used to be thought that circumcision benefited personal hygiene, allowing the penis to be kept properly clean. As a result, a high percentage of adult males were circumcised. In some places, the procedure is no longer routine, but in the U.S., 6 out of 10 babies have their foreskin removed (*above*).

A circumcised penis is often less sensitive than with an intact foreskin. The mucous membrane of the glans (tip) becomes toughened, or cornified, through constant contact with underwear and loses its soft moistness. Some men say that they find this an advantage, since the reduction in sensitivity allows them to prolong their lovemaking, but there is little evidence that circumcision has a significant effect on a man's sexual response. The important thing is that you can still enjoy better sex whether or not you have been circumcised.

SURGICAL PROCEDURES

There is a current vogue for penis enlargement. Surgery to inject the base of the penis with silicone – or fat sucked from the abdominal wall – aims to increase the weight of the organ by about 1 oz (30 g) and increase the width of its base. Another operation intended to increase penis length by 50 percent involves separating out the root of the penis from the pelvis, bringing it forward, and restabilizing it by stitching it to the pubic bone.

If you are seriously worried about the size of your penis, it is worth seeking advice from your doctor.

ERECTIONS

Several emotional, physical, and hormonal signals trigger erections.

The penis is made up of inflatable cylinders of erectile tissue – there are two upper corpora cavernosa (cavernous bodies) that lie side by side and a lower corpus spongiosum (spongy body) containing the urethra. The urethra is the tube through which urine flows from the bladder out of your body.

At the tip of the penis, the corpus spongiosum expands to form the bulky glans penis, or helmet. The other end of the corpus spongiosum thickens

contract during orgasm to propel semen forward. The root of the penis is often overlooked as an erogenous zone when making love (*see pp. 50–51*), and exercising the pelvic floor muscles can lead to more intense sensations during orgasm (*see p. 57*).

WHAT IS AN ERECTION?

The enlarging and stiffening of the penis in response to sexual arousal is caused when the spongy tissues in the penis

FLACCID PENIS

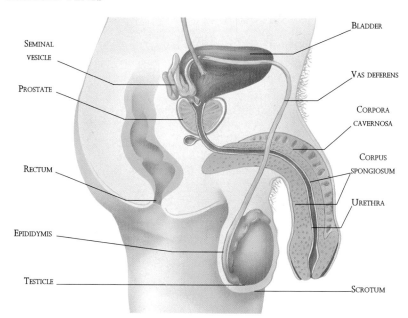

SEMINAL VESICLE

PROSTATE

RECTUM

EPIDIDYMIS

TESTICLE

BLADDER

VAS DEFERENS

CORPORA CAVERNOSA

CORPUS SPONGIOSUM

URETHRA

SCROTUM

behind the scrotum to form the root of the penis. The base of the penis is connected to muscles (bulbospongiosus and ischiocavernosus) that allow it to swing upright into an erection and

swell as their blood supply arteries dilate. The corpora cavernosa and the corpus spongiosum all fill with blood and become rigid. This causes a rapid increase in pressure locally, which compresses the outlet veins and prevents the blood from draining away. While the blood remains trapped, the erection is maintained.

CONTROLLING YOUR ERECTIONS

Most men reach climax and ejaculate too quickly at some time in their life. For 50 percent of males, this happens when first making love with a new partner. This is an indication of extreme arousal and should be seen as a compliment by your partner.

There are several ways you can enjoy better sex if you feel that you often come too soon. These include:

◆ Bringing your partner to the point of orgasm through direct clitoral stimulation during foreplay – penetration can then occur before, during, or after her orgasm.

◆ Wearing a condom to reduce sensations, and using cream containing a local anesthetic to numb the tip of the penis.

◆ Tensing your buttock muscles while thrusting to block nerve signals from your penis.

◆ Thinking of something other than sex in order to psychologically lessen your arousal.

◆ Gently pulling your testicles back down into the scrotum when they rise up to the base of the penis when fully aroused – be careful not to twist them.

◆ Using the "squeeze" technique – you or your partner squeezes your penis between the thumb and two fingers just below the tip, where the glans joins the shaft. Squeeze firmly for five seconds, then wait for a minute before resuming sex.

ERECT PENIS

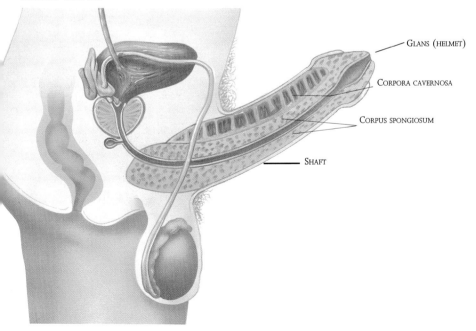

GLANS (HELMET)

CORPORA CAVERNOSA

CORPUS SPONGIOSUM

SHAFT

THE TESTICLES

In each of these sperm "factories," 50,000 sperm are made per second.

The testicles, or testes, are responsible for producing sperm and the male sex hormone testosterone. Each testicle (or testis) is shaped like an egg and measures ¾–1 in (2–2.8 cm) across and about 1½ in (4–4.25 cm) in length. It is quite normal for one testicle to hang lower than the other. Each testis contains up to a thousand tightly coiled seminiferous tubules where the sperm are made.

Sperm and testicular secretions drain through a network of vessels into the epididymis. This is a 20-ft (6-m) long, tightly coiled tube attached to the top of each testicle. As sperm pass through the epididymis, they start to mature and become motile, which means that they start to move spontaneously and independently. The epididymis on each side leads into a vas deferens.

To help sperm and testosterone production, the testicles need to be about 7–12.5°F (4–7°C) cooler than overall body temperature. This is why the scrotum holds the testicles outside the body. If you are concerned about sperm production, you can:
◆ Avoid hot baths – even bathing in water at 109°F (43°C) for half an hour a day can affect your testicles.
◆ Wear loose-fitting, cotton underpants rather than tight, synthetic ones – this prevents the buildup of heat and electrostatic fields, both of which can damage testicular health.
◆ Splash your testicles with cold water two or three times per day.

INSIDE THE TESTICLES

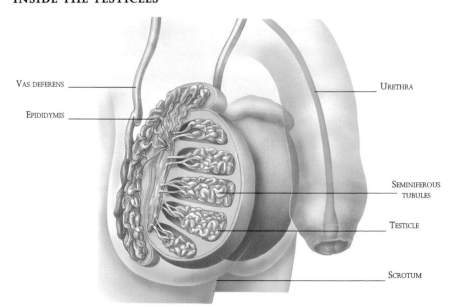

VAS DEFERENS

EPIDIDYMIS

URETHRA

SEMINIFEROUS TUBULES

TESTICLE

SCROTUM

EXAMINING YOUR TESTICLES

You can contribute to your sexual health by examining your testicles regularly – about once a month. Testicular cancer can often be caught at an early stage when it is still treatable and the prospects for full recovery are much better. The best place to do this self-exam is in the bath or shower, where the scrotum is warm and relaxed.

Each testicle should feel soft and smooth, like a hard-boiled egg without its shell. If you notice any lumps, swelling, irregularities, tenderness, or abnormal hardness, consult your doctor immediately.

◆ Hold each testicle in turn between the thumb and fingertips of both of your hands.
◆ Slowly and gently bring the thumb and fingertips of one hand together while relaxing the fingertips of the other.
◆ Alternate this movement several times so that you learn what your testicles feel like and can assess each one.

VASECTOMY AND SEX

During a vasectomy, the two narrow muscular tubes known as the vasa deferentia are cut and tied (*above right*). These tubes connect each epididymis to the penis and act as sperm storage sites. During orgasm, they contract to pump secretions up into the penis.

After vasectomy, the ejaculated volume of semen – the thick, whitish fluid that would normally contain the sperm produced in the testicles – lessens by approximately 5–10 percent, but this is usually unnoticed.

POINTS OF INTERRUPTION

TUBES ARE CUT AND TIED AT THESE POINTS DURING A VASECTOMY

One recent study found that up to 40 percent of men who had undergone a vasectomy had noticed a change in their usual ejaculation pattern. About 1 man in 6 reported a more rapid ejaculation, while nearly 1 in 4 reported that ejaculation was delayed. (The pregnancy failure rate for the operation is 1 in 2,000.)

After a vasectomy, blood levels of testosterone go up, since testicular secretions are being absorbed back into the body instead of being ejaculated. These increased levels of testosterone in the blood may be responsible for an initial increase in sex drive. However, one study found that about 10 years after the operation, testosterone levels were lower than average – this may result in a subsequent decrease in sex drive in some men. This finding is under further investigation.

THE PROSTATE GLAND

Few men know precisely where their prostate gland is – or what it does.

The prostate gland is hidden away between the bladder and the penis, wrapped around the urinary tube (urethra). In a healthy man, the prostate weighs about ¾ oz (20 g) and is the size and shape of a large chestnut. It is made up of millions of tiny glands that produce a milky fluid.

calcium, copper, fructose (a sugar), and substances – such as spermine and putrescine – that give semen its characteristic smell.
◆ Directs semen outward during ejaculation so that it does not pass backward into the bladder.
◆ Makes hormone-like substances

LOCATING THE PROSTATE GLAND

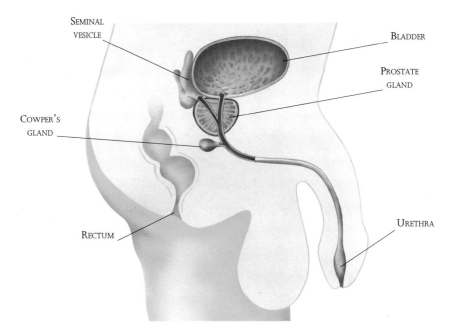

SEMINAL
VESICLE

BLADDER

PROSTATE
GLAND

COWPER'S
GLAND

RECTUM

URETHRA

The prostate gland:
◆ Secretes fluids that make up 30–40 percent of semen volume.
◆ Adds more than 30 different nutrients to semen. These include zinc, amino acids, citric acid, vitamin C, vitamin B_{12}, sulfur, magnesium,

known as prostaglandins.
◆ Makes enzymes (such as acid phosphatase) that regulate semen clotting.

If prostate secretions are not released regularly through intercourse or masturbation, they build up to trigger feelings of fullness, increased sex drive, and wet dreams (*see p. 28*). Just beyond the prostate are two tiny structures known as Cowper's glands. These provide early lubrication during sexual arousal, often seen as glistening drops of liquid at the end of the penis before ejaculation occurs.

PROSTATE PROBLEMS

The prostate gland can interfere with your enjoyment of sex if it becomes inflamed (prostatitis) or enlarged (benign prostatic hypertrophy). Inflammation causes pain, especially when urinating or making love. Enlargement can mean that the urethra is compressed as it passes through the gland, interfering with urine flow. This can also cause problems with erections and ejaculation. You should always report any problem with urination to your doctor.

PROSTATE HEALTH

Simple changes in your lifestyle can make a world of difference to the health of your prostate gland:

◆ Eat more oriental-style foods, such as soy products, rice, kohlrabi, and Chinese cabbage. These contain natural plant hormones that help to decrease the effect of dihydro-testosterone – a breakdown product of the male hormone, testosterone – which has been linked with unwanted enlargement of the prostate gland. Men who follow this type of diet are also less likely to develop prostate cancer.

◆ Eat a low-fat diet containing little red meat – eat fish and skinless chicken instead. This has been shown to significantly decrease the risk of developing prostate cancer.

◆ Eat at least five servings of fresh fruit or vegetables per day. This will help to provide essential vitamins and minerals such as antioxidants and zinc that are vital for the healthy functioning of your prostate.

◆ Increase your intake of nuts, seeds, and whole grains, especially rye products (including extracts of rye pollen or saw palmetto if you are already experiencing prostate problems).

Research shows that substances contained in these products can decrease congestion, swelling, and inflammation of the prostate gland. This, in turn, helps to reduce such problems as prostatitis, which can interfere with both erections and urination.

◆ Consider taking dietary supplements containing vitamins C and E, as well as betacarotene and zinc to protect the health of your prostate.

◆ Avoid taking strenuous exercise when your bladder is full. Chemical irritation and inflammation of the prostate can occur when urine refluxes up into the gland. This may lead to prostatitis.

◆ Try yoga positions designed to decongest the prostate and pelvic areas (*see pp. 44–45*).

THE WOMAN'S BREASTS

Breasts have long been worshiped as symbols of female sexuality, fertility, and motherhood.

Female breasts come in many shapes and sizes, and they are a unique symbol of each woman's individual sexuality. Many women feel that their breasts are either too big or too small. It is important that you make your partner feel at ease with her body so that she can relax when she is naked with you – and enjoy each sexual encounter to the full.

Some women prefer gentle, delicate handling of their breasts, while others seem to be less sensitive in this area of their anatomy and prefer slightly rougher fondling, squeezing, and sucking. You will have to experiment to see which sensations your partner prefers.

Sex therapists investigating the sexual responses of women have found that about one woman in a hundred can actually achieve an orgasm by having her breasts and nipples stroked and fondled, even with no other sexual stimulation.

Bear in mind, however, that some women find any type of breast stimulation uncomfortable – usually just before their period, when the breasts can become very tender. In one study, only about half the women surveyed claimed to enjoy having their breasts fondled as a part of foreplay.

SHAPE AND COLOR CHANGES

Caressing, kissing, and fondling a woman's breasts as part of foreplay will cause them to swell slightly as she becomes aroused. As sexual excitement mounts, the nipples become erect, and the colored area around them, known as the areolae, may look swollen or appear to be bruised. At the plateau stage of excitement (*see also pp. 56–57*), which occurs just before she reaches orgasm, a pink flush may appear on her chest.

After orgasm, during the resolution stage, the pink flush clears, and her breasts, nipples, and areolae all resume their normal appearance once again.

EFFECTS OF CHILDBEARING

How a woman's breasts respond to sexual stimulation also depends to some degree on whether or not she has had a child and if she breast-fed the infant. Women who have never been pregnant will usually have pale, rose-colored nipples and areolae. After pregnancy, however, the nipples and the surrounding areas usually become noticeably darker. If a woman has breast-fed her child for a significant length of time, her breasts will not go through the same pattern of swelling as they did before childbirth. It appears that breast-feeding causes changes in the supply, circulation, and drainage of blood from the area of the chest wall, with the result that swelling and shape changes usually no longer occur during sexual arousal.

FEMALE ANATOMY

Being familiar with female genitalia will help you to become a better lover.

The outer part of the woman's reproductive tract, consisting of the clitoris and vagina, is actively involved in sexual intercourse. The internal organs (uterus, ovaries, and Fallopian tubes), are tucked away in the pelvis, but they still contribute to the pleasure of orgasm when waves of muscular contractions pass through them.

THE VAGINA

This space is normally about 3¼ in (8 cm) long. The corrugated front and back walls are in contact and form an H-shape in cross section that can expand to accommodate any size of penis. During sexual arousal, vaginal tissues become increasingly engorged with blood, changing in color from rosy pink to purplish red. The upper

two-thirds of the vagina lengthens and balloons outward, and the uterus is pulled up, producing an intense desire for penetration. At the same time, the entrance to the vagina pouts downward, and the lower third of the vagina thickens to grip the penis more tightly.

THE HYMEN

In childhood, the entrance to the vagina is protected by a thin membrane – the hymen – which often tears naturally during such physical activities as gymnastics or cycling. Sometimes, it tears only during a woman's first penetrative sexual encounter and may bleed slightly. However, due to the fragility of the hymen, a lack of pain and bleeding is equally common.

GENITALIA

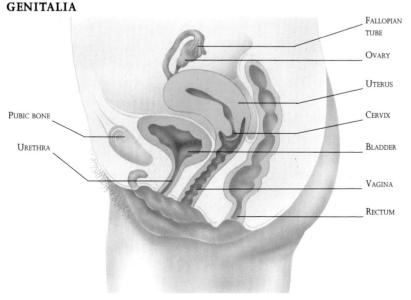

FALLOPIAN TUBE

OVARY

UTERUS

CERVIX

BLADDER

VAGINA

RECTUM

PUBIC BONE

URETHRA

THE CLITORIS

The most sensitive female erogenous zone is the clitoris. It is ¾–1¼ in (2–3 cm) long and sits above the entrance to the vagina, where the lips meet at the top. Usually the organ is retracted, but when a woman is sexually aroused, it swells, lengthens, and becomes more prominent. During sexual intercourse, penile thrusting only indirectly stimulates the clitoris by stretching surrounding tissues. Most women require some form of direct stimulation of the clitoris as well – with the fingers, the tongue, or the end of an erect penis – to reach orgasm.

STIMULATING THE CLITORIS

Direct stimulation of the clitoris should be gentle and sensitive, with plenty of lubrication – dry or rough handling will produce unpleasant sensations. Most women find it uncomfortable for the clitoris to be stroked upward, since this can irritate the urethral opening beneath. When stroking or licking the clitoris, use a downward action (from the pubic hair toward the vagina). Try making gentle around-and-around motions to one side of the clitoris using two or three fingers. Alternatively, use your palm or the tips of your fingers to stimulate the entire area using a slow circular motion. Most women also enjoy being penetrated by one finger while their clitoris is being stimulated – but make sure that your fingernails are short when trying this.

THE G-SPOT

Perhaps the most controversial – and sought after – erogenous zone, the G-spot bears the initial of its discoverer, Ernst Grafenberg. It is said to be a button-like area of tissue on the front wall of the vagina that swells during sexual arousal. When the G-spot is stimulated and gently pressed, a rapid and intense orgasm is experienced.

Some researchers believe that the G-spot may be the female equivalent of the prostate gland (*see pp. 20–21*), since it is situated near the neck of the bladder.

Hunting for the G-spot during foreplay is an excellent way for both of you to become sexually aroused. Your partner should lie on her back while you lie on your side next to her.

◆ Insert your lubricated index finger 2–3 in (5–7.5 cm) into her vagina. Making a beckoning gesture with your finger should bring your fingertip up against the front vaginal wall, where you can feel the underlying hardness of her pubic bone.

◆ Press your fingertip firmly, but gently, up against this bone and move it around and around in a small circle, slowly working upward or downward. If she suddenly feels a strange, pleasant and intense sensation, you've found it!

THE G-SPOT

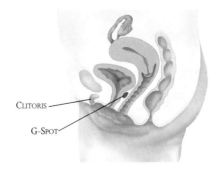

CLITORIS

G-SPOT

HELPING YOURSELF

There are many things you can do to improve your own sexual health. You may already be aware that you need to lose some excess weight, tone up your abdominal muscles, or cut back on your drinking or smoking. You may also need to make sure you are getting an adequate supply of essential vitamins and minerals – through paying close attention to your diet and, in some cases, through taking supplements.

Better sex revolves around learning to be at ease with yourself, your body, and its sexual responses. By getting to know your body, what it looks like, and how it feels and reacts, you will start to become more sexually confident.

Virtually everybody experiences their first orgasm through masturbation. This self-exploration plays an important role in your progress toward sexual intercourse. For adults, masturbation can act as a map, guiding you toward better, more fulfilling sex.

Many men are as unadventurous when masturbating as they are when making love. It is an activity that is usually carried out speedily and somewhat furtively, rather than as a form of self-discovery. Masturbation affords an ideal time to experiment with new fantasies, to try different ways of stimulating your body, and to learn how to delay orgasm to reach a fantastic climax.

MASTURBATION

Learning about your own body and its sexual responses is one step on the path to better sex.

The word masturbation simply means the manipulation and stimulation of your own or your partner's sexual organs to produce an orgasm. It is an entirely natural, healthy activity that is essential for men who are not in a regular sexual relationship. It is also frequently enjoyed by couples who consider autoeroticism – or mutual masturbation – to be an integral part of their foreplay.

Surveys reveal that 80 percent of men masturbate regularly – 13 percent of males masturbate more than three times per week, 25 percent between one and three times per week, and 15 percent masturbate two or three times per month. Men who aren't sexually active and who don't masturbate will eventually experience a wet dream.

A man usually masturbates by moving one or both of his hands up and down the shaft of his penis. He may also rub the fold of skin – the frenulum – that attaches the foreskin to the tip of the penis underneath.

To make masturbation more pleasurable, instead of concentrating solely on your penis, let your hands stroke your chest, abdomen, thighs, scrotum, perineum, buttocks, and even your anus. As you become more aroused and approach ejaculation, slow down to prolong these pleasurable feelings. Learn to bring yourself to the edge of orgasm, and then retreat, several times. Also, try experimenting with lotions,

massage oils, and talcum powder to experience different sensations on different parts of your body. You may find it arousing to watch yourself by positioning a full-length mirror beside the bed.

If you have sex infrequently, you may find that you ejaculate quickly once inside your partner. One way to help overcome this is to masturbate a few hours before lovemaking.

To masturbate your partner, stimulate her clitoris with a lubricated finger or vibrator. In addition, caress her breasts, nipples, inner thighs, and the lips of her vagina to increase arousal. For some women, masturbation is the only way they can achieve an orgasm – roughly 50 percent of women are unable to reach a climax during penetrative sex.

WET DREAMS

Most men have up to five erections while they are asleep. These usually last around 30 minutes each and are often present on waking. If you have not ejaculated for several days, a build-up of secretions may trigger a nocturnal emission in which ejaculation occurs during sleep. This is perfectly normal and is nature's way of getting rid of sperm that are past their expiration date. Wet dreams are most common during adolescence. Men who enjoy a regular sex life or who masturbate regularly are less likely to experience them.

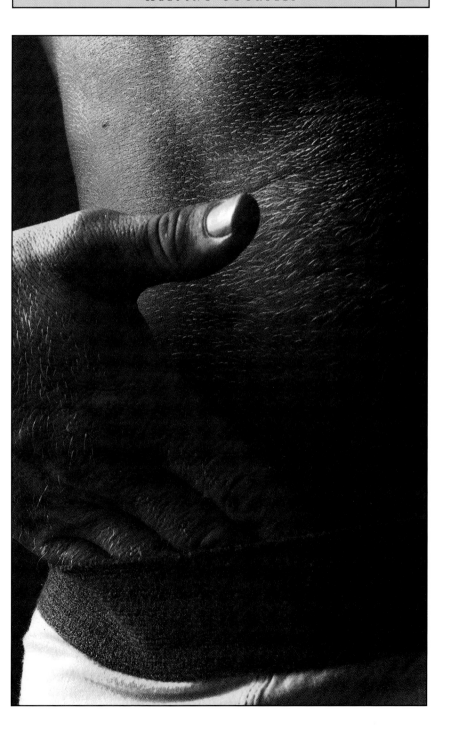

GETTING FIT

Caring about your body and well-being is essential for improved vitality and increased self-esteem.

Unless you feel confident about your own body, both physically and mentally, and about your general level of fitness, it will be difficult for you to enjoy better sex. Taking several simple steps may be all you need to enhance your sexual potential.

LOSE EXCESS WEIGHT

This is best done slowly and steadily through a combination of healthy eating, reduced fat intake, and increased physical activity. Sex itself can burn calories and will help you to lose weight, especially if it is vigorous and prolonged.

EXERCISES TO IMPROVE SEXUAL STAMINA

The types of exercise that are most beneficial to your sexual health and stamina are generally those that are "high impact" since these build up your strength and endurance. Such examples include jogging, running, tennis, and racquetball. Other exercises such as brisk walking, cycling, and swimming are also generally beneficial.

If you prefer to work out in a gym, follow a comprehensive program that strengthens the muscles in your back, abdomen, upper arms, buttocks, thighs, and calves. Ideally, the exercise should be of moderate intensity and performed on a regular basis – three or four times a week for 20 to 40 minutes. Once you are fit, try to find time to do some form of exercise every day as a maintenance program.

TIPS FOR IMPROVING YOUR GENERAL LEVEL OF FITNESS

If you find it difficult to build an exercise plan into your busy day, try the following:
◆ Walk short distances instead of taking the car, bus, or train.
◆ Walk up stairs instead of using the elevator or escalator.
◆ Try to take a short walk around the block at least once a day – for example, during your lunch break.
◆ If you can't go out, try running up and down the stairs a few times.
◆ Do sit-ups, jog in place, or use a home exercise machine when watching television in the evening.
◆ Walk briskly whenever possible.

MEDICAL CHECK-UPS

Have an annual physical to make sure that your blood pressure and urine are both normal and that your heart and kidneys are functioning properly. Supplement this yearly check-up with a monthly testicular self-examination (*see pp.18–19*). If you develop any symptoms that last longer than a week or so, always seek professional medical advice. It is difficult to enjoy a better sex life if you are worried about – or suffering from – poor health, so you should always take your own health seriously.

CHANGING YOUR LIFESTYLE

Enjoying a better sex life means following a healthy, active daily routine.

In order to be able to enjoy a better, more fulfilling sex life, you may have to be prepared to make a few adjustments to your lifestyle. Most of the following information regarding health matters is well known, but you may not be aware of precisely how these behaviors can affect your sexual performance.

ALCOHOL CONSUMPTION

Although alcohol may enhance desire, it also takes away the ability to perform sexually. Not many people realize that drinking more than the recommended safe maximum on a regular basis can have serious long-term effects on your sexuality, however.

Alcohol slows testosterone secretion and hastens its conversion to estrogen in the liver. In the short term, this can lead to decreased sex drive, impotence, and a low sperm count – about 40 percent of male infertility has been blamed on moderate alcohol intake. Continued overindulgence in alcohol may eventually lead to shrinking testicles, a reduction in penis size, and a loss of pubic hair. Specific advice on the safe limits for alcohol consumption is difficult to give, since the amount that may trigger these types of problems varies from person to person, depending on how your metabolism handles alcohol and how much exercise you get. In general, an intake of 2-3 alcoholic drinks per day is unlikely to have any harmful effects on your health.

CIGARETTE SMOKING

Smoking tobacco generates massive amounts of free radicals – harmful by-products of metabolism. These interfere with testicular function and can damage sperm. Free radicals are absorbed by dietary antioxidants such as vitamins C and E, as well as by betacarotene, which is converted to vitamin A in the liver.

Smokers use up their vitamin C supplies quite rapidly, and men who smoke 20 or more cigarettes every day have blood levels of vitamin C that are up to 40 percent lower than those of non-smokers leading an otherwise similar lifestyle. Smokers also have sperm counts that are 17 percent lower than nonsmokers, reduced sperm motility, and a greater percentage of abnormal sperm.

A daily intake of 200 mg of vitamin C can improve the sperm count of smokers by as much as 24 percent, sperm motility by up to 18 percent, and the number of sperm still alive 24 hours after ejaculation by 23 percent. Sperm are also less likely to clump together. (It is only individual sperm that are capable of entering and fertilizing an egg.)

A measurable improvement in the health of sperm seems to start within approximately one week of increasing your vitamin C intake.

THE EFFECTS OF STRESS

You cannot hope to enjoy better sex if you are stressed out, tired, and

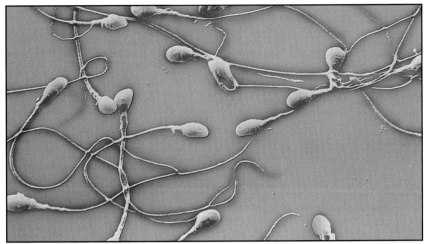

USING COLOR-ENHANCED
imaging techniques, normal sperm appear like individual tadpoles (*above*). Abnormal sperm are clumped together and have misshapen head capsules or tails (*below*).

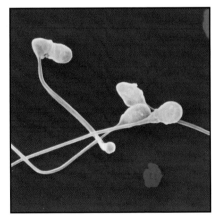

overworked, and these are all commonly recognized factors that contribute to a low sex drive and poor erections.

For the sake of your general health and of your sex life, try to avoid stressful situations. Make time for relaxing and pursuing the leisure activities you enjoy, and make sure that you get plenty of sleep.

When stressful situations are unavoidable, try avoiding caffeine and nicotine, since these mimic the body's stress responses and add to your stress load. During times of stress, increase your intake of vitamins B and C. Both of these are rapidly used up in the metabolic reactions associated with stress.

SYMPTOMS OF STRESS

If you suffer from any of these symptoms, see your doctor or get some help to reduce your stress levels:
- tiredness and difficulty sleeping
- sweating or flushing
- rapid pulse or palpitations
- dizziness or faintness
- trembling, numbness, or pins and needles
- chest pain
- diarrhea
- overwhelming anxiety or panic
- fear of rejection or failure
- moodiness
- increased aggression or anger
- loss of sex drive.

NUTRITION

Improving your diet can boost your energy level and improve your sex life.

Lack of certain vitamins or minerals may affect your health and lead to such common symptoms as fatigue, irritability, difficulty in concentrating, and a lowered sex drive. Research suggests that many men are commonly deficient in zinc, iron, magnesium, calcium, and vitamins B_1, B_2, B_6, C, and D. Use this chart to discover which foods are the best sources of these vital vitamins and minerals.

In general, aim to eat a healthy diet:
◆ Eat several small meals throughout the day instead of large meals three times a day.
◆ Always have breakfast.
◆ Eat a wide variety of different foods.
◆ Eat fresh, unprocessed food.
◆ Eat more fresh fruits and vegetables.
◆ Eat less red meat and more oily fish.
◆ Drink either skim or low-fat milk.
◆ Cut back on your intake of salt, sugar, and junk foods.

NUTRIENT
Vitamin A
Vitamin B complex
Folic acid
Vitamin C
Vitamin E
Iron
Calcium
Magnesium
Chromium
Copper
Manganese
Iodine
Selenium
Zinc

GOOD SOURCES	BENEFICIAL EFFECTS
Cod-liver or halibut oil, butter, milk, egg yolks, liver, fruit	Helps to regulate growth, sexual development, and reproduction.
Meat, liver, yeast extract, fortified cereals, bread, Brussels sprouts, cauliflower, oranges, melons	B_3 (niacin) is said to be an aphrodisiac; B_6 regulates sex hormone function.
Spinach, broccoli, Brussels sprouts, parsley, soybean products, whole grains	Important for the cells lining the seminiferous tubules in the testicles.
Fruit, especially citrus, potatoes, tomatoes, vegetables	Essential for sperm and prostate health and for fighting off infections.
Wheat germ oil, avocados, margarine, whole grain cereals, seeds, nuts, oily fish	Protects sex hormones from breaking down; helps sex drive and fertility.
Red meat, liver, bread, flour, egg yolks, dried fruit, vegetables	Helps in production of hemoglobin and maintaining energy levels.
Dairy products, fish, canned sardines, bread, green vegetables	Needed for muscle contractions during orgasm.
Soybeans, nuts, whole grains, seafood, dairy products, bananas, dark green leafy vegetables	Helps to maintain a healthy balance of male sex hormones.
Egg yolks, red meat, cheese, fruit, honey, whole grains	Maintains sex drive, sperm count, and fertility.
Shrimp, shellfish, olives, nuts, seeds, legumes, whole grains	Maintains sex drive, sperm count, and fertility.
Black tea leaves, whole grains, nuts, seeds, fruit, eggs, seafood, vegetables	Maintains sex drive, sperm count, and fertility.
Seafood, seaweed, iodized salt, milk	Maintains stamina and sex drive.
Broccoli, mushrooms, cabbage, onions, garlic, whole grains, nuts, seafood	Needed for the production of sex hormones and for maintaining sex drive, sperm count, and fertility.
Red meat, seafood (especially oysters), whole grains, legumes, eggs, cheese	Essential for male sexual maturity, testicular function, and potency. Many men become zinc deficient since each ejaculation contains about 5 mg zinc – which is about one third of the daily requirement.

SAFE SEX

It is every individual's responsibility to practice safe sex.

You cannot fully enjoy sex if you are worried about the possibility of unwanted pregnancy or sexually transmitted disease. Condoms offer a good level of protection against pregnancy and substantially reduce the risk of contracting, or passing on to your partner, gonorrhea, NSU (chlamydia), trichomonas, herpes, hepatitis B, and HIV. A spermicide, such as nonoxynol-9, adds to this level of protection, since it can kill some infections outright in the same way that it kills sperm.

If used carefully (in conjunction with spermicidal jelly), male condoms have an excellent record for safeguarding against unwanted pregnancy – with only about a 2 percent failure rate. This rate, however, rises to 25 percent with careless use. Condoms used "dry" are most likely to burst during vigorous sex. But take care not to use mineral-based oils (for example, baby oil,

petroleum jelly, and some spermicidal creams) with latex condoms, since these substances dissolve rubber and can reduce condom strength by up to 95 percent within just 15 minutes.

VARIETY

Using condoms can even add to your enjoyment of sex. Modern condoms are made from the highest-quality pre-lubricated latex or an ultra-thin polyurethane for improved sensitivity. Worldwide, there are two standard widths 2.05 in and 1.9 in (52 mm and 49 mm) and several different lengths. A variety of different shapes is available to provide optimum fit and sensitivity (*opposite*). When choosing a condom, check to see if it is labeled "not to be used as a barrier." If it does carry this, or a similar warning, use it only for fun – in other words, do not rely on it for protection against unwanted pregnancy or disease.

TIPS WHEN USING CONDOMS

◆ Check the expiration date. Avoid any genital to genital contact until the condom is in place. Open the foil packet carefully so the condom is not damaged. Do not use a condom if the wrapping is cracked or torn.
◆ Encourage your partner to put the condom on you as part of foreplay – this can prevent loss of spontaneity and passion at a vital moment.
◆ Squeeze the nipple of the condom to release air before rolling it onto your fully erect penis using your other hand. Roll it right down to the base of your penis.
◆ Immediately after ejaculation, grasp the penis and condom near the base and hold firmly while you withdraw from your partner – don't stay inside until you deflate, since the condom may slip off of a nonerect penis.
◆ Wrap the used condom in a tissue and dispose of it hygienically.

TYPES OF CONDOMS

Flared condoms are for those with a large penile tip (glans).

Contoured ones give a snug fit below the head of the penis and are flared over the glans for improved sensitivity.

Extra-thin condoms are supposed to improve sensitivity for the penis (you can also try those made from polyurethane).

Snugger fitting are for men with a narrow penis.

Extra-wide are for men with a thicker penis.

COLORED
AND
FLAVORED

Ribbed or dotted condoms claim to increase friction, and may improve clitoral stimulation for your partner. However, use plenty of water-based lubricant in order to prevent soreness. Do not use these condoms for oral sex, since they may make your partner's mouth sore.

Nonspermicide lubricant condoms should be used if either partner is allergic to spermicides.

Polyurethane condoms are for increased sensitivity. You can apply your own choice of lubricants, and

TEATLESS

they are also recommended for those who are allergic to latex.

Colored are fun types for variety.

Flavored are for safer oral sex.

Teatless brands have no nipple to get in the way during oral sex.

Novelty condoms glow in the dark (fundoms), and some have sport or "racing" stripes.

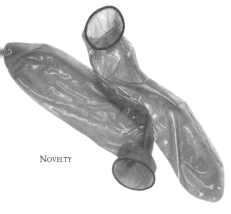

NOVELTY

EXERCISES FOR YOUR SEX LIFE 1

Tone up your pelvic muscles to add some spark to your sex life.

The movements involved when making love are different from those used in everyday life. You can bring about a real improvement in your sexual performance by toning up your pelvic muscles and by maximizing the flexibility and mobility of the associated tendons. Releasing tension and congestion in the pelvis, abdomen, lower back, and legs will improve the flow of your sexual energy and help you to enjoy a better sex life.

These following exercises are best performed after a bath or shower, when you are warm and your muscles are relaxed.

PELVIC GYRATIONS
◆ Stand with your feet apart and your knees slightly bent. Place your hands on your hips, stick your bottom out, and rotate your pelvis slowly in a clockwise direction.
◆ Continue rotating your pelvis around in a complete circle that is as wide as possible.
◆ Continue these pelvic gyrations for 1–2 minutes, and then gyrate your pelvis around in a counterclockwise direction for another 1–2 minutes.

BACKWARD LEG LIFTS

◆ Lie face down on the ground with your abdomen tucked in, your pelvis tilted forward, and your hips pressed to the floor.

◆ Rest your head on your right arm and stretch the other arm out in front, palm facing downward. Breathe in and flex your right foot.

◆ As you breathe out, raise your right foot as high as you can, keeping your toes pointing downward, your back straight, and your hips on the floor. Hold for 2 seconds. Lower your leg back to the ground.

◆ Repeat the exercise 10 times with each leg.

PELVIC LIFTS

◆ Lie on the floor, pushing the small of your back down so that it touches the ground. Bend your knees, place your feet flat on the floor and your arms by your side, palms down.

◆ As you breathe out, roll your bottom upward so that it lifts off the floor. Clench your buttocks and inner thigh muscles.

◆ As you breathe in, roll your bottom down to the floor again and relax your clenched muscles.

◆ Repeat these pelvic lifts 5 to 10 times.

EXERCISES FOR YOUR SEX LIFE 2

*The following exercises will help
strengthen your lower back, buttocks,
and thigh muscles.*

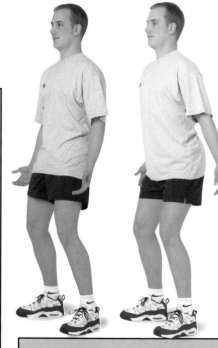

ABDOMINAL CRUNCHES
◆ Lie on your back with your knees bent and the soles of your feet flat on the floor. Keep your back straight and in contact with the floor.
◆ Rest your palms on the front of your thighs. While your head is still on the ground, pull in your abdominal muscles and clench them tightly.
◆ Keeping your abdominal muscles clenched, breathe out and slowly lift your head and shoulders from the floor, so that your palms move toward your knees.
◆ Move smoothly and keep your lower back on the floor and your chin up.
◆ Hold this position for 2 seconds. Keeping your abdominal muscles clenched, lower your shoulders to the floor as you breathe in again.
◆ Repeat this exercise 10 times as a continuous movement.

PELVIC TILTS
◆ Stand with your feet apart and your knees slightly bent.
◆ Hold your arms by your side with your palms facing forward. Breathe in deeply and pull your pelvis back.
◆ As you breathe out, let your pelvis rock forward and your arms, hands, and genitals move upward.
◆ Repeat these pelvic tilts in a smooth, continuous motion for about 2–5 minutes.

LEG SQUATS

◆ Stand with your feet apart and your knees slightly bent. Keep your back straight, abdomen tucked in and pelvis tilted forward. Rest your hands comfortably on your hips.

◆ As you breathe in, bend at the knees and hips and squat down slowly until your thighs are parallel to the floor.

◆ Keep your back straight and body aligned, your heels on the floor and your weight centered directly over your ankles.

◆ As you breathe out, slowly stand up until you reach your original position. Repeat 10 times.

BACK RAISES

◆ Lie face down with your feet together and arms stretched out on the floor in front of you. Keep your forehead on the ground throughout.

◆ Pull your abdominal muscles in and push your pelvis into the floor. Clench your buttocks and breathe in slowly.

◆ As you breathe out, raise your right arm and left leg to a comfortable height. Do not jerk in an attempt to get higher. Hold for a count of 3 seconds, then lower your limbs to the floor as you breathe in again. Rest for a count of 3 seconds.

◆ Repeat this exercise using your left arm and right leg, for a total of 10 times per side.

EXERCISES FOR YOUR SEX LIFE 3

Simple and undemanding, these exercises will not be a problem regardless of your general level of fitness.

Just as regular exercise tones up muscles in your body to improve your stamina, strength, and flexibility, the exercises in this book are designed to improve the muscles used when making love. As well as strengthening your back, abdomen, arms and legs, they will have a beneficial effect on the small muscles in your pelvis that are involved in your sexual response.

There is no muscle in the penis itself, but you should see some improvement in your sexual performance by strengthening the muscles in your pelvic floor. This group of muscles stretches from your pubic bone to your coccyx. Exercising your pelvic floor muscles will help to strengthen the muscles that support your penis when it is erect. It will also help to tone up the accessory muscles that contract during orgasm.

Some researchers claim that by exercising these muscles, you will be able to exert more control over your orgasm and that you may even be able to reduce the amount of fluid ejaculated each time. Some men have been able to train themselves to have a dry orgasm – one without any ejaculation at all (see p. 59). By retaining at least some semen, it is possible for a man to enjoy a series of multiple orgasms in quick succession.

Even if this technique does not work for you, you will still gain immense benefit from exercising these muscles. Most people experience a more intense orgasm after just a few weeks of intimate workouts.

IDENTIFYING YOUR PELVIC FLOOR MUSCLES

At first, it can be difficult to isolate the pelvic floor muscles you need to exercise. The easiest way to identify them is while you are urinating. When you are in mid-stream, concentrate on stopping the flow completely. Practice this every time you go to the bathroom until you are able to squeeze these muscles whenever you want – when sitting, walking, or lying down, for example. Once you can do this, start exercising them regularly.

The simple routine for these exercises is as follows:
◆ Start with 10 quick squeezes, holding each one for a count of three seconds. Repeat this two or three times each day.
◆ Build up your exercise routine to 20 quick squeezes at a time, holding each for a count of three seconds. After about a month of this, add in 5 long, slow squeezes after your 20 quick squeezes.
◆ For the long, slow squeezes, clench your pelvic floor muscles while counting to 10 and then hold the muscles clenched for another 10 seconds.
◆ When you are comfortable with these, build up to 10 long, slow squeezes after each series of 20 quick squeezes.

Once you can identify your pelvic floor muscles, you can practice this exercise routine almost as a reflex.

SQUEEZING YOUR PROSTATE GLAND
This exercise strengthens the anal muscles and massages the prostate. This helps to keep the gland well toned to increase the strength of contractions within the gland during orgasm. It may also increase secretion of hormones and produce a natural high.
◆ Sit, stand, or lie, whichever you find most comfortable.
◆ Squeeze your anal muscles tightly and hold for as long as possible.
◆ Relax for a minute and repeat for as many times as is comfortable.

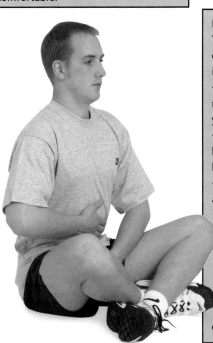

RELEASING TENSION
This exercise helps to open energy channels and free blockages that may interfere with sperm production.
◆ Sit cross-legged on the floor and relax your body. Make sure to keep your spine straight.
◆ Gently cup your scrotum in your left hand and place the palm of your right hand on your abdomen.
◆ Massage the left-hand side of your abdomen for 2 minutes using a circular motion.
◆ Change hands so that you are cupping your scrotum with your right hand and repeat the abdominal massage – this time on the right-hand side – for 2 minutes.
◆ Then sit and relax while breathing deeply and evenly for 1 minute.

EXERCISES FOR YOUR SEX LIFE 4

Several traditional yoga exercises can give sex a new dimension.

Many yoga positions are designed to increase your body's overall suppleness and to strengthen important muscle groups. The one shown here may also decrease pelvic congestion and encourage drainage of the prostate gland.

CHATURANGA DANDASANA (FOUR-LIMB STICK) POSTURE

◆ Lean forward, bending at the knees and hips, and place your hands on the floor in front of your feet.

◆ Keeping your hands flat on the floor, walk your legs backward and stiffen your body, slowly lowering yourself to the floor in a push-up position.

◆ Keep your elbows tucked in to the sides of your body. Your weight should be centered over your toes and the palms of your hands.

◆ If this is difficult at first, lower your knees to the floor. Hold this position for as long as you find it comfortable.

URDHVA MUKA SVANASANA (UPWARD FACING DOG) POSTURE

◆ Continuing from the Chaturanga Dandasana position, lower your knees to the floor, straightening your toes behind you.

◆ Lift your body up by straightening your arms. Your weight should be on your hands and the tops of your feet.

◆ Pull your head back to look upward. Do not let your back sag or your shoulders hunch up.

◆ Your posture is correct if you do not feel any discomfort in the area of your lower back. Hold the position for as long as it is comfortable.

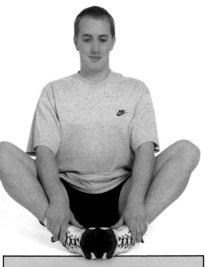

BADDHA KONASANA (BOUND ANGLE OR COBBLER'S) POSTURE

◆ Sit on the floor and bring the soles of your feet together, pulling your heels up as close as possible to your genitals. If the adductor muscles in your upper thighs are tight, adopting this position may be quite difficult on the first try.

◆ Keep your back and shoulders straight and your head up. Try to let your knees sink down toward the floor while making sure the soles of your feet are still flat against each other.

◆ Hold your ankles and relax for as long as it feels comfortable. This position drains congestion away from your pelvis.

Note: Shoe repairers in India habitually sit in this position – hence one of its English names of Cobbler's Posture – and they are known for rarely having problems with their prostate gland.

UPAVISHTA KONASANA (SEATED ANGLE) POSTURE

◆ Sit on the floor. Separate your legs as wide as you possibly can. As you breathe in, reach forward and try to grasp your toes. If you cannot reach them, grasp your ankles or shins instead.

◆ Lift your chest upward, keeping your back straight, and look upward.

◆ Breathing out, bend forward as far as possible, pushing the back of your knees into the floor so that your knees point straight up.

◆ Keep your thigh and pelvic floor muscles contracted.

◆ Hold this position for 10 seconds.

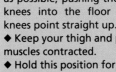

BOOSTING YOUR SELF-CONFIDENCE

To enjoy your sex life, you need to have confidence in yourself.

It is easy – and very destructive – to compare yourself unfavorably with others. If you are worried about your looks, the size of your penis, or how you measure up as a lover, it will be difficult to relax when making love. However, if you feel at ease with yourself and your body, you will exude an aura that attracts people.

LOSING NEGATIVE THOUGHTS
Take off all your clothes in front of a full-length mirror. Some relaxing music may help, since this exercise takes a little time.
◆ Study your face and body thoroughly.
◆ Turn sideways and study your profile and rear view, too. (A triple mirror, or another long mirror placed at an angle will help.) Write down a list of your good points – for example, *my skin has a healthy glow; my hair is thick and healthy; my abdomen is firm and flat; my penis is normal sized.* Don't be afraid to be honest.
◆ Follow this by writing down a list of all the negative things about your body – such as *my skin looks sallow; my hair needs cutting; I'm developing a potbelly; my penis is too small.*
◆ Having studied your body, put on some loose, comfortable clothes that you feel relaxed in and think about your personality, emotions, and sexual experiences. Write down all the positive things you can think of – *I have a great sense of humor; women seem to like talking to me; I have*

no difficulty finding a girlfriend; I am caring and loyal; I am sensitive to a woman's needs.
◆ Finally, write down all the negative thoughts that come into your head – *my relationships don't seem to last; I am not sexually desirable; I'm not very good in bed.*

Now look at the lists containing the negative physical and emotional statements, and turn these into positive ones (and include any solutions). For example:
◆ *My skin looks sallow* becomes: *I will improve the appearance of my skin through diet, exercise, and having a series of facials.*
◆ *I'm beginning to develop a potbelly* becomes: *I will firm up my abdomen through a combination of diet and exercise.*
◆ *My penis is too small* becomes: *my penis is perfectly normal (see pp. 14–15).*
◆ *I am not sexually desirable* becomes: *I am sexually desirable.*
◆ *My relationships don't seem to last* becomes: *my relationships will last.*
◆ *I'm not very good in bed* becomes: *I will become good in bed and enjoy sex.*

Don't underestimate the power of positive thought. Every day, read the list of good things you wrote down about yourself and recite the new positive statements aloud to reinforce your motivation. This will begin to change your negative patterns of thought into positive thoughts that will help you shed your inhibitions and become a more confident, better lover.

SEX AND YOUR PARTNER

Making love is a physical expression of the depth of feeling you and your partner have for each other. These feelings should permeate your whole relationship and set the scene for the physical act of love.

By allowing sexual tension to build up slowly, perhaps over a period of hours, your enjoyment of the physical act that follows will be that much greater. Luxuriate in the feelings of sexual arousal; fantasize if you wish, and learn to share your fantasies with your partner.

If you are dining out before making love, pay particular attention to the food you order. Neither too much or too little food is conducive to better sex. Similarly, go easy on the wine – sharing a bottle between two is fine. More than this may increase your desire but ruin your performance and make you sleepy.

Take time to set the scene. Make sure that wherever you choose to make love is private. If there are children in the house, lock the bedroom door if necessary. Many couples who have been together for a while tend to undress themselves and then meet in bed naked. Why miss out on the pleasures of slowly undressing each other while kissing and fondling? Mood music and candlelight may also add to the sensuality of the experience.

BETTER FOREPLAY

Touching and caressing can help bring you and your partner to full sexual arousal.

When a man is said to be good in bed, it usually means he is good at foreplay. A lack of good (or enough) foreplay may mean that your erection does not become fully hard and you may not be able to ejaculate; for a woman, bad foreplay may mean that she stays dry, making penetration uncomfortable and orgasm less likely. The keys to better foreplay are the major and minor erogenous zones.

The major erogenous zones are sensitive areas of skin covered with a high density of nerve endings. When these areas are stimulated in the right way, they can trigger sensual feelings and sexual arousal. Erogenous zones are important because they provide a short cut to arousal when making love.

During foreplay, these sensitive areas of skin can be stimulated by gentle stroking, licking, blowing, or nibbling. The minor erogenous zones vary from person to person, so while gently blowing on the back of the neck is a turn-on for some, others prefer having their ear lobes nibbled. Working out where your zones are – and discovering your partner's – is all part of the fun of making love.

MAJOR EROGENOUS ZONES

Your major erogenous zones include all the obvious sites. In men, the major zones are the lips, nipples, penis, and scrotum. In women, they are the lips, buttocks, breasts, nipples, and external genitals, including the clitoris and vaginal entrance.

MINOR EROGENOUS ZONES

Everyone has secondary erogenous zones, including the back of the neck, the eyelids, ears and ear lobes, plus the soft skin at the top of the inner thighs. Most people are also sensitive around the anus, but be careful – although this sensation is pleasurable for some, others find it distinctly uncomfortable.

COMMUNICATION

It is important that both partners communicate with each other when making love. This can mean whispering, talking, breathing words of encouragement, or a message to stop and try something else. Your processes of sexual arousal are finely tuned – you need to be able to interpret each other's moans of pleasure to know when you are doing something just right. Equally, you need warning when you are doing something less than perfectly.

Let your partner know where you like being touched – either by telling her or by guiding her hands or fingers. It may be difficult or embarrassing to say things such as "harder," "faster," "softer," "yes, just like that," or "slower," but these messages are vital for better sex, and can even be a turn-on. Some people can get these messages across simply by the tone and pitch of their moans. If you are unsure of what your partner is trying to tell you, ask. A whispered "Is that good?" or "Do you want me to stop?" will usually get you back on the right track.

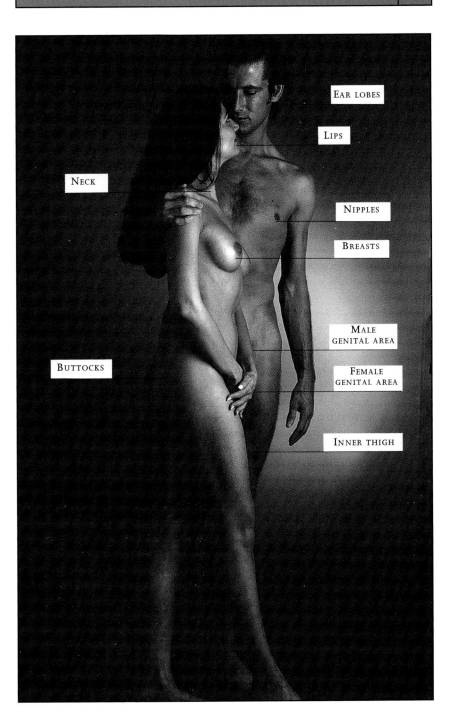

EAR LOBES

LIPS

NECK

NIPPLES

BREASTS

MALE
GENITAL AREA

BUTTOCKS

FEMALE
GENITAL AREA

INNER THIGH

SENSUAL MASSAGE

This is the best way to create the mood for an exciting sexual encounter.

When indulging in this erotic form of foreplay, use a specially formulated massage lotion – one containing diluted sensual aromatherapy oils is ideal. If you are using rubber barrier contraceptives, do not use a mineral-based oil (*see p. 36*).

♦ Your partner should lie naked, face down. Smooth massage lotion onto her back using large circling movements to warm her skin. Continue with slow, gliding, fluid motions that follow the natural curves and contours of her body. If you find a muscle that seems knotted or tense, gently knead that area. Keep asking her what she likes and be alert for appreciative noises.

♦ When you have finished her back, cover it with a towel and move downwards to her buttocks and the back of one leg – try to keep the other leg covered. Massage the leg right down to her feet, then cover that leg and concentrate on the other.

♦ Ask her to turn over so you can massage her front. Gently stroke her chest and shoulders, avoiding her breasts at first. Slowly work your way down to her abdomen and then the front of her legs. Start to tease her a little by drawing your fingers up close to her breasts and genitals then pulling away. When she starts to become aroused, massage her breasts, gently tweaking the nipples to see if she likes this. Finally, concentrate on massaging the insides of her thighs. This is a powerful erogenous zone in most women. Try alternating firm

movements with feathery ones, slowly working your way towards her genitals. If you want to experiment with oral sex, remember not to get any massage oil or cream on the area – or switch to using an edible lubricant (for example, olive oil, whipped cream, yogurt, or flavored gels). The clitoris is the most sensitve part of the female genitals and this area should only be stimulated after she is fully aroused. Use soft, slow and gentle strokes with plenty of lubrication.

♦ By now, you should both be aroused, and you need to decide whether you want your partner to enjoy an orgasm on her own or whether to progress to full penetration. However, a session of mutual masturbation can be equally enjoyable for both of you.

AROMATHERAPY OILS
The following oils are thought to have sensual properties:

Ylang ylang is an intensely sweet scent that stimulates erotic thoughts.

Tuberose, with its rich, heavy aroma, boosts feelings of sensuality.

Jasmine's powerful, seductive effect helps you to relax and dispels stress.

Rose otto's erotic scent, full of oriental mystery encourages closeness.

Rose maroc has an earthy smell with warm, dark, erotic undertones.

Sandalwood's sweet, rich scent has relaxing, seductive properties.

MASSAGE TECHNIQUES

Massaging your partner helps to heighten sexual arousal for both of you.

Body massage can be the best way to give and receive pleasure as a prelude to lovemaking. Different techniques should be used to enhance stimulation and enjoyment. Try to change the strength, rhythm, and style of stroke from soft, languid touches to firm, invigorating movements. Start slowly, and build up gradually to include as much of the body as possible so that no part feels left out. This allows the whole body to become responsive to pleasure and erotic feelings.

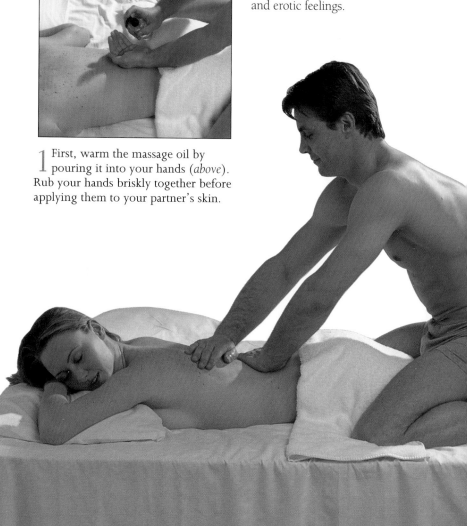

1 First, warm the massage oil by pouring it into your hands (*above*). Rub your hands briskly together before applying them to your partner's skin.

3 Some people like the light pummelling sensations more than the softer, gliding ones. It is a very light, springy movement, not the heavy karate-chop used by professionals to release tight muscle knots. Experiment to find out which strokes you like best.

2 Start the massage with sweeping, circular motions on each side of your partner's spine (*above*). Begin high up, near the shoulder blades, and work slowly down toward the buttocks, keeping your hands relaxed and taking care not to drag the skin.

4 The fleshier parts of the body are more likely to be areas of tension caused by knotted muscles. Gently knead these areas using your fingertips, thumbs, or knuckles, rhythmically squeezing and releasing the flesh (*below*).

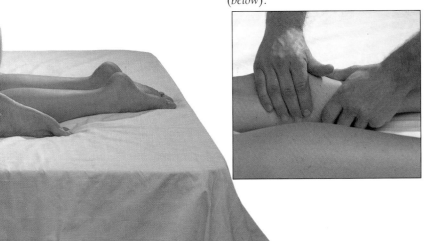

AROUSAL AND ORGASM

An insight into the different stages of your orgasm.

Having an orgasm involves going through four distinct stages.

STAGE 1: EXCITEMENT

Stimuli from a number of sources (psychological, tactile, and visual) reach your brain to arouse your sexual interest. This triggers an erection. Arousal can easily be interrupted by distractions such as stressful thoughts or the telephone ringing.

The excitement phase in men is short, usually less than 5 minutes. In women, it can last 15 or even 30 minutes.

STAGE 2: PLATEAU

As arousal becomes more advanced, hormones released from the adrenal glands speed up your heart rate and your blood pressure rises. Your breathing becomes more rapid. Many people develop a flush that starts on the lower abdomen or chest and spreads to the neck, face, and limbs. It is also common to start perspiring. Your nipples become erect and your scrotal skin thickens and contracts. The testicles are drawn up toward the base of the penis and may increase in volume by 50 percent due to congestion with blood. The tip of the penis (glans) also swells to improve friction.

The plateau phase lasts about 15 minutes in men. In women, it is usually shorter – around 5 minutes – since it produces an intense desire for penetration. The thrusting of the penis increases a woman's arousal level gradually until she reaches orgasm.

MALE AND FEMALE ORGASM

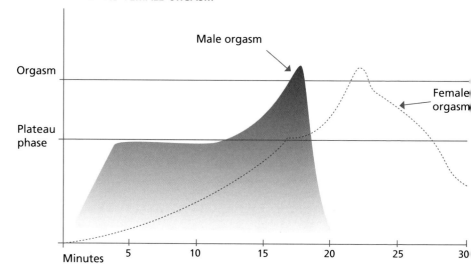

Stage 3: Orgasm

If stimulation is adequate, the physical effects noted during the plateau phase become more intense, and you experience a feeling that orgasm is inevitable. In women, orgasm can be delayed by external distractions. However, men experience a "point of no return"; the feeling of inevitability is due to hormonal secretions, and ejaculation is unstoppable.

Orgasm itself is an intensely pleasurable sensation, variously described as emanating in the head, the penis, testicles, or everywhere. It is accompanied in men by a varying number of major and minor muscle contractions. Nerve impulses spread via the pudendal nerves and cause rhythmic, wavelike contractions of the pelvic floor muscles and sometimes of the thigh muscles as well. Contractions of muscles lining the vasa deferentia propel sperm up from the testicles and out through the penis.

During orgasm, a number of brain chemicals are released that intensify the feeling of arousal and pleasure. Your heart rate and blood pressure peak, and hyperventilation is common. You may also be aware that your rectal sphincter muscles are contracting. Involuntary vocalizations often occur as pleasurable sensations wash through you.

Orgasm produces between four and eight major muscle contractions. They start at intervals of about 0.8 seconds and then reduce in both strength and frequency. Orgasms usually last anywhere between 3 and 10 seconds; they rarely last longer than about 15 seconds.

Stage 4: Resolution

After orgasm, a period of resolution follows, when heart rate, blood pressure, and genital blood flow all return to their normal state. Nerve impulses trigger the relaxation of the muscles lining the reproductive tract, and the arteries supplying blood to the penis close down. The muscle fibers lining the blood spaces in the penis contract, pushing blood out of the spongy tissues. This relieves the pressure on the outlet veins so that pooled blood can quickly drain away, encouraging penile flaccidity.

Resolution occurs rapidly in men – over the space of a few minutes – as long as orgasm has been reached. If, however, the plateau phase does not result in orgasm, resolution may take several hours to complete. This may cause a dragging sensation in the loins and testicles, which may be uncomfortable.

In women, resolution is more gradual, with breasts, nipples, and labia sometimes taking as long as 30 minutes to return to their normal size.

The refractory period

After orgasm, men experience an absolute refractory period when further orgasm is impossible. This is probably related to the high levels of epinephrine in the blood and the activation of inhibitory centers in the brain. In young men, this period is often only a few minutes; after middle age, it often lasts much longer. Women do not experience this, and so multiple orgasms can occur more easily. Some women experience a series of mini-orgasms lasting up to a minute.

PROLONGED AND MULTIPLE ORGASM

Learning to control your orgasm helps you to prolong your pleasure.

By deliberately delaying your climax during the plateau stage of orgasm (*see pp. 56–57*), you can enhance – and usually prolong – the sensations felt when you ejaculate. Delay allows a buildup of male hormonal secretions, and you will have more intense contractions as a result. Some people can train themselves to experience orgasm while retaining some, or all, of their seminal fluid – in other words, only partially ejaculating (*far right*) – so that they can go on to enjoy multiple orgasms.

EXPERIENCING MULTIPLE ORGASMS

It is a well-known fact that some women can enjoy multiple orgasms

one after another in a series extending over a period of a minute or more. What is less widely appreciated is that some men regularly experience multiple orgasms too – and that it is possible to teach yourself the technique.

For a man, having a multiple orgasm simply means that you experience more than one climax during a single love-making session, which could stretch over a period of an entire evening. Many males can do this – but it is also possible for a man to climax more than once in quick succession during a single act of love-making. Although this is not common, it is known to occur, especially in adolescent males who experience only a

MULTIPLE ORGASMS

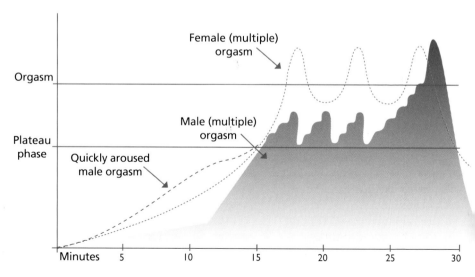

PROLONGED ORGASMS

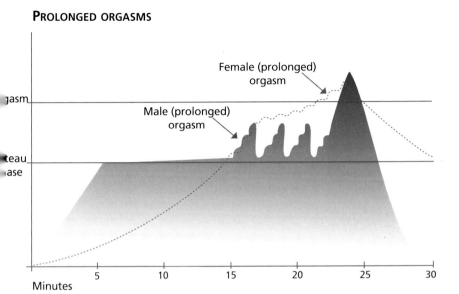

Female (prolonged) orgasm

Male (prolonged) orgasm

gasm

teau
ase

Minutes
5   10   15   20   25   30

brief refractory period (*see pp. 56–57*).

Multiple orgasms are a sought-after goal by both men and women. This is because each subsequent orgasm is usually more intense and satisfying than the one preceding it. This is particularly true in males, as muscular contractions in the pelvis become more pronounced as less seminal fluid remains in the tubes for expulsion. Being capable of a multiple orgasm also prolongs the length of time you and your partner can make love, which increases the pleasure for both of you.

LEARNING THE TECHNIQUE

There are two techniques you can try to reach orgasm without ejaculation.

◆ Use the squeeze technique to prolong your erections (*see p. 16*).
◆ When you feel that orgasm is imminent, stop completely and relax deeply while clamping down on your pelvic floor muscles (as if you are trying to stop yourself urinating in mid-flow – *see p. 42*). One sex researcher described interviewing a man who claimed to have come three times in the space of just ten minutes by using this technique of not fully ejaculating until the final time.

By practicing these two techniques, you should learn to time your reactions so that, as you relax, the sensations of orgasm sweep through you without full ejaculation taking place.

HAVING A BETTER ORGASM

Helping your partner to reach orgasm is one measure of being a better lover.

It usually takes a woman much longer to become aroused than it does a man, and your partner may need a lot of clitoral stimulation to reach her climax. This requires patience and skill on your part during foreplay (*see pp. 50–51*).

Few women are able to achieve orgasm during the thrusting action of penetrative intercourse itself, so, if your partner is unable to come this way, remember that this is perfectly normal and it is no reflection on you. However, don't give up without first experimenting with some of the arousal techniques given in this book.

About 8 percent of adult women have never experienced an orgasm, even when masturbating. Many more do not experience one until they have been sexually active for several years.

foreplay or penetration, a technique known as the bridge maneuver can help her to climax during intercourse.

To perform this maneuver, you need to stimulate your partner's clitoris gently with your lubricated fingers or tongue during foreplay. When she is about to come, she should let you know. You should then immediately insert your penis into her vagina and *slowly* move it up and down. This is often enough to trigger her orgasm.

At first, she may need you to continue direct stimulation of her clitoris with your hand or fingers, too. This is most easily done if you both lie side by side, facing the same way. Your partner should bend the knee of her uppermost leg and drop it backward so that it rests over you as you lie

THE BRIDGE MANEUVER
If your partner does
not easily reach
an orgasm
during

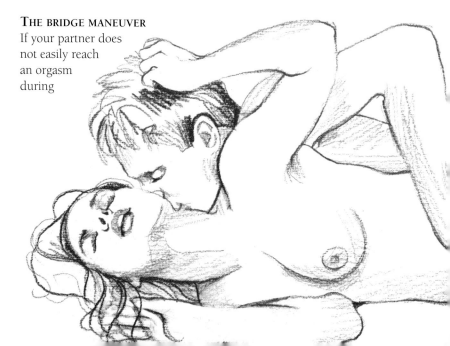

behind her. You should then be able to reach over and gently stimulate her clitoris with one hand while inserting your penis from behind.

Some women prefer penetration to occur only at the point of climax; others prefer to feel their partner's penis inside them right from the start.

OTHER TECHNIQUES

You can delay your orgasm by prolonging foreplay and the excitement phase of orgasm to maintain sexual tension for as long as possible. Once penetration has taken place, slow down or stop thrusting altogether when you feel that your ejaculation is imminent – but before it becomes inevitable. This technique will also help your partner to reach a higher level of arousal.

If your partner takes a long time to reach her climax, your delaying tactics will increase her chances of having an orgasm during penetrative sex, especially if you can reach down and provide gentle clitoral stimulation, too. According to recent research studies, foreplay that is prolonged for a minimum of 20 minutes and thrusting

sustained for at least 15 minutes (including some rest periods to catch your breath and prevent soreness due to friction) is usually sufficient for the majority of women to reach climax.

SIMULTANEOUS ORGASMS

Many couples consider reaching orgasm together to be the highest measure of better sex. To achieve simultaneous orgasms, you need to know your partner's sexual responses as well as your own in order to gauge how quickly her excitement is mounting.

Once you know that she is approaching orgasm, you can let yourself go. Some men find that their partner's muscular contractions during orgasm is enough to trigger their own climax, and vice versa.

AFTERPLAY

For women, afterplay is just as important as foreplay and penetration. Holding her close, caressing her gently, and telling her how much you enjoyed your lovemaking will help her to feel comfortable, respected, safe, and loved.

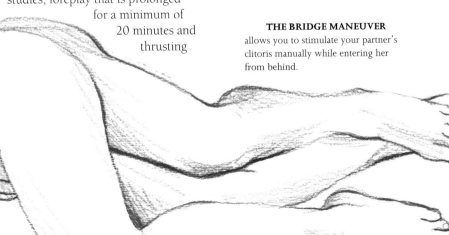

THE BRIDGE MANEUVER allows you to stimulate your partner's clitoris manually while entering her from behind.

ORAL SEX

Experimenting with different ways of giving pleasure makes you a better lover.

One of the most enjoyable and sexually arousing activities for both partners, oral sex is a very special form of intimacy in which you surrender yourself totally to giving or receiving pleasure. There are two types, one for each sex: fellatio is the licking and sucking of the penis; cunnilingus is the licking and sucking of the clitoris and vagina. The smell and taste of each other's secretions are designed by nature to be highly sexually arousing. But beware of eating garlic or too many spicy foods, since these may interfere with your natural smells.

BE PREPARED

Safer oral sex involves using a barrier against sexually transmittable diseases. For the man, this means wearing a condom (*see pp. 36–37*). For the woman, this means placing a dental dam – a large, thin square of latex – over her genitals and licking and stimulating her through this. Many couples also like to shower or bathe together before indulging in oral sex so that they can cleanse each other as an erotic part of foreplay.

FELLATIO

Most men enjoy receiving fellatio, although it is important that you always let the woman control the depth to which she takes your penis into her mouth. If you thrust into her,

it can produce an unpleasant gagging sensation, which is a real turn-off.

Not all women are aware that the most sensitive part of a man's penis is the helmet at the top, especially the ridge. At the back of the helmet is a fold of skin, the frenulum, which is also

RECEIVING FELLATIO IS HIGHLY AROUSING for most men. To be enjoyable for both of the people involved, each must put their complete trust in the other.

very sensitive. Flicking the tongue along the ridge and frenulum is known as the butterfly technique. You may need to tell your partner the areas where you would like her to concentrate. Let her know, too, whether you want her to vary the speed and pressure with which she manipulates your penis.

Warn your partner when you are about to come (if the signs are not obvious) so that it is her decision whether or not to swallow your semen if you are not wearing a condom.

CUNNILINGUS

The soft moistness of the tongue suits a woman's delicate genitals and is less likely to cause irritation and dryness than using your fingers. Your partner is the best person to tell or show you exactly what she likes, since every woman is different.

Experiment together and respond to her sighs, moans, and pelvic movements to discover how your tongue can thrill her. Don't just concentrate on the clitoris: gently nuzzle or lick the whole genital area, including her inner thighs. Change the speed and pressure of your movements and try slipping a finger inside her vagina. The secret of good cunnilingus is to keep changing what you do, not pursuing one stroke for too long – unless she indicates that this is what she wants you to do.

The physical signs of arousal to watch for are the swelling and pouting of the vaginal lips, hardening and lengthening of the clitoris, and general engorgement of the area. One word of caution, however – never blow into the vagina (especially if your partner is pregnant). This may be dangerous, since it potentially can force air into the bloodstream through the uterine lining.

SIXTY-NINE

The sixty-nine position involves lying on your back (or side) with your partner lying on top (or to one side), head to toe, so that you both have access to each other's genitals. Not everybody likes this position, since it can be difficult to concentrate on receiving pleasure while trying to give it at the same time.

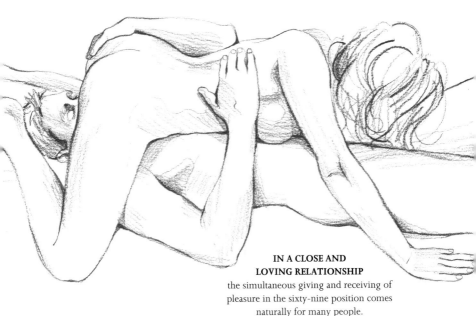

IN A CLOSE AND LOVING RELATIONSHIP the simultaneous giving and receiving of pleasure in the sixty-nine position comes naturally for many people.

GOOD SEX POSITIONS

Varying your sex positions can open up new ways of giving and receiving pleasure.

Man-on-top positions let you control your thrusting and the speed with which you reach orgasm. This is useful either for hastening your climax (say, when making love for a second or third time, when your erection is less hard) or for delaying orgasm for a more satisfying experience.

Woman-on-top positions give control to your partner. By moving up and down and varying the speed and depth of penetration, she can set her own pace and reach orgasm more easily. If you feel you are closer to reaching orgasm than she is, give a signal so that she can stop moving for a few moments.

OPEN AND SHUT CASE

(for better clitoral stimulation)
With you lying on your back, your partner sits on top of you, leans back, and balances herself by resting both hands on your knees. You then have full access to her breasts, and as she moves up and down, you can stimulate her clitoris with one hand.
Note: Use a lubricant to avoid any unpleasant friction.

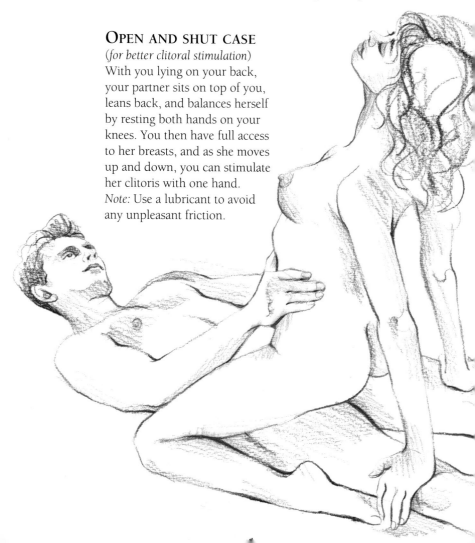

JOYRIDER

(good position for the second time around)
This position lets your penis stimulate
your partner's clitoris and outer vaginal
lips, even if your penis is not fully
erect. While you kneel down on
the bed, your partner gently
lowers herself on top, sitting on
your thighs. She then controls the
rate of movement and the depth of
penetration. You also have the
freedom to move up and down at
the knees and to thrust your
pelvis forward if you wish.

GOOD SEX POSITIONS

NUTCRACKER

(for the tightest grasp of the penis)

Your partner lies down overhanging the bed. Kneel down between her legs and push forward to enter her. She then reaches down with one hand and makes a "V" with her fingers, so that a finger slips down on either side of the base of your penis. She squeezes her fingers together to grip you firmly as you move in and out. This position also lets her stimulate her clitoris with her own hand and gives you access to stimulate her breasts and clitoris, too.

HURDLE POSITION
(*for better control and deeper penetration*)
The woman lies on her back, her buttocks on a pillow, and bends one knee up to her chest. With you on top, you support your weight with one hand on the bed or the side of your partner, and the other hand placed on her bent knee, pushing it down. This lets you thrust deeper and increases the sensation of tightness.
Note: Try to avoid this position in the second half of pregnancy because it will put unwanted pressure on the womb.

UPSIDE-DOWN MISSIONARY POSITION
(*for better clitoral stimulation*)
With you lying on your back and your partner lying on top of you, she brings her legs inside yours for a snug fit. By keeping her legs tightly closed, she can increase the friction between your penis and her vagina and clitoris.

GOOD SEX POSITIONS

WINDMILL POSITION
(for a different slant)
With your partner lying on her back, you lie across her at an angle of about 45 degrees. When you enter her, you will be stimulating the side of her vagina, producing an unusual sensation. Try rotating slowly toward or away from your partner to find the most pleasurable position for you both.

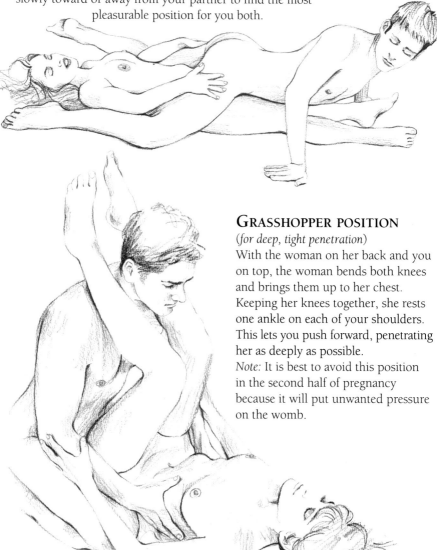

GRASSHOPPER POSITION
(for deep, tight penetration)
With the woman on her back and you on top, the woman bends both knees and brings them up to her chest. Keeping her knees together, she rests one ankle on each of your shoulders. This lets you push forward, penetrating her as deeply as possible.
Note: It is best to avoid this position in the second half of pregnancy because it will put unwanted pressure on the womb.

REAR ENTRY POSITION

(*to stimulate the G-spot*)

Sexual penetration from behind often gives the most satisfying orgasms for women, since lovemaking in this position thins and stretches the front wall of the vagina where the G-spot is said to be situated. Your partner leans forward, either supporting herself on the side of the bed or with her arms back holding your waist. Grasp her waist or buttocks so that you can penetrate her deeply from behind. As an alternative, try kneeling on a bed rather than standing.

These positions allow full stroke movement, deep penetration, and also give you access to your partner's clitoris and breasts. Another position good for stimulating the G-spot is lying curled up, side by side, with your front to her back, like spoons resting together.

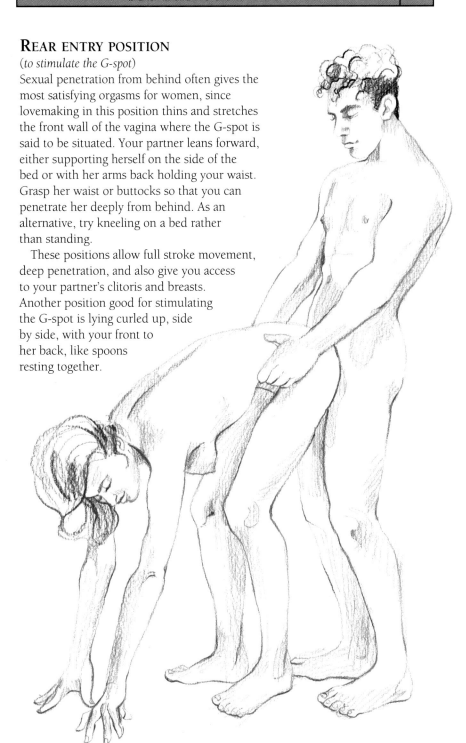

ADVANCED TECHNIQUES

Different sexual activities can breathe new life into your lovemaking.

Once you have discovered where each other's major and minor erogenous zones are located (*see pp. 50–51*), you can start experimenting with some more advanced techniques for stimulating them. Use a feather to make long, soft, tickly sweeps down the insides of each other's thighs, for example, or use a vibrator on some of the less obvious places of each other's anatomy, such as the backs of the knees. Also, the base of your penis is a very sensitive area, yet it is frequently overlooked when making love.

PLAYING HOT AND COLD

Some men are really turned on by having their scrotum or the back of their neck stimulated, alternately, with warm and then icy cold water. This technique can also be used on each other's nipples. Try sucking an ice cube before kissing your partner somewhere intimate.

PERINEAL PRESSURE

Many positions allow your partner to reach down and press your perineum (the area found between your scrotum and anus) firmly during intercourse – this can be highly arousing when you are thrusting, for example. The pressure adds to that of secretions building up inside you. Pressing and then stopping can help you to approach and then delay orgasm several times in order to produce a more intense and satisfying climax.

ANAL STIMULATION

The anus is extremely sensitive, especially around the rim nearest the vagina or testicles. Try stimulating these areas with a well-lubricated finger. Some people enjoy "rimming," in which the tip of the finger is slipped inside (keep your fingernails short). Always wash your hands thoroughly afterward, and before you touch your partner's genitals again.

PROSTATE MASSAGE

The prostate gland is thought to be the equivalent of a woman's G-spot (*see p. 25*). If the prostate is stimulated before or during intercourse, it can produce an especially intense orgasm.

Either you or your partner can stimulate the gland by inserting a lubricated index finger into your anus. The prostate will be felt behind the front wall of the rectum as a firm, walnut-sized swelling (*see pp. 20–21*). Gently massaging this gland will produce high feelings of arousal. In a few men it can trigger a rapid orgasm. (If the gland is excessively tender, seek medical advice, since it may be inflamed, a condition known as prostatitis.)

Prostate massage is not as messy as you might think, since the lower rectum is usually empty. Wash your hands after a prostate massage, and before touching your partner's genitals – bowel bacteria can trigger cystitis and vaginal infection in women.

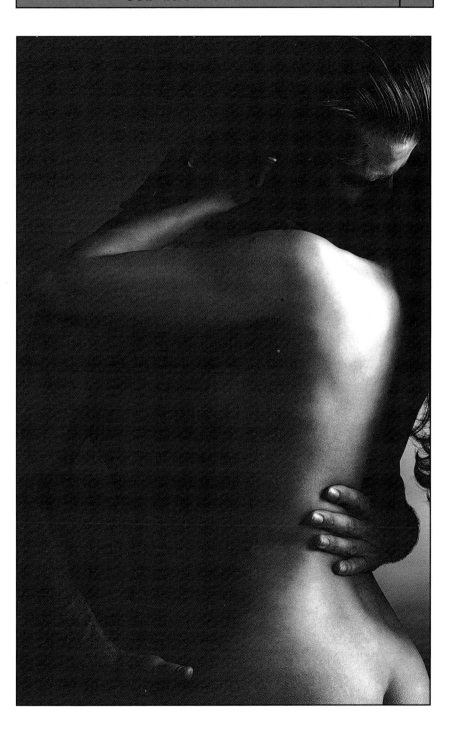

SPONTANEITY AND FANTASY

Don't let your sex life become routine – make love as the mood moves you.

Many couples make love only in the bedroom and only at night, but why not make love on the kitchen table, in the bathroom, on the living room floor, or sprawled up the stairs? Some couple may prefer to lock the doors and pull the curtains to shut themselves off from the outside world. For others, half the excitement comes from the risk of being discovered making love outdoors – in a park or on the beach, perhaps.

IN THE BATH
Making love in a warm bath with plenty of bubbles – or in the shower – can also be erotic. This gives you the opportunity to soap each other down in intimate places. The best position for sex in the bath is with the woman sitting on top of you. She can use the sides of the tub to help lift herself up and down. Be careful, however, if using condoms, as they can be ineffective.

USING A MIRROR
Increase your excitement by using a large mirror, which can lead to better sex. A mirror adds a voyeuristic element, allowing you to watch your partner's facial responses and body. You can also witness penetration itself, which may be highly arousing for both of you.

FANTASIZING
Approximately 95 percent of men and women regularly entertain erotic thoughts to arouse them. There are four main types of fantasy: those involving past, present, or imaginary lovers; scenes indicating sexual power and irresistibility; imagining different settings, practices, and positions; and thoughts of submission or dominance.

You are more likely to fantasize about somebody you know than about a famous personality, but only about a third of people surveyed said that they fantasize about the person they are with at the time.

Some couples find that sharing their fantasies, or acting them out, is a turn-on, while others prefer to keep their intimate images private. If your relationship is strong, try sharing fantasies as a way of keeping each other's interest alive.

FLOTATION TANKS
You can experiment with producing new and erotic fantasies in a flotation tank. A tank is filled with a warm, highly salted solution. You are suspended without sensory distractions from light, sound, or temperature stimuli, and the effects of gravity are significantly reduced. This encourages the creative side of your brain to go into overdrive since it no longer has to filter millions of signals every second from the outside world. Floating is good for your health, too, as it lowers both blood pressure and stress levels. Most large urban areas have a float center – check your telephone directory.

APHRODISIACS

Substances to stimulate sexual desire have been used for centuries.

There are several types of food, drinks, drugs, scents, and other devices that are believed to stimulate sexual desire and improve sexual potency. Throughout history, several of the more unusual aphrodisiacs have included boiled goats' testicles, powdered beetles, and rhinoceros horn.

One of the most widely used of all the aphrodisiacs is the old favorite, alcohol, which, in moderate amounts, reduces inhibitions and stimulates desire. If consumed to excess, however, it has the opposite effect – it can make you feel sick to your stomach or even render you unconscious.

You may want to experiment with some of the following and see how they work:

◆ Oysters – said to mimic the smell, flavor, and texture of the female genitals. Even if you do not agree, they are still a rich source of zinc (*see pp. 34–35*).

◆ Pitted fresh litchi fruits (or the similar fruit, rambutan) – inserting your tongue in the hole left by the pit is said to mimic the sensations of cunnilingus (*see pp. 62–63*).

◆ Caviar – the sharp, salty, fishy flavor of the fish eggs is thought to be reminiscent of musky female juices.

◆ Nutmeg – a pinch of powdered nutmeg mashed into the flesh of an avocado, which is then chilled for 24 hours, is supposed to have powerful aphrodisiac properties for men. An amphetamine derivative in the nutmeg (methylenedioxyamphetamine) reacts with a chemical in the avocado

(bromocriptine) to produce a sexual stimulant that is effective only in men.

◆ Figs – thought to resemble the female external genitalia. Both the Greeks and the Romans considered figs to be a powerful aphrodisiac.

◆ Ginger – this root is believed in Chinese medicine to be a powerful aphrodisiac. Its hot, pungent, spicy flavor has a warming and loosening effect on the system in general and on the libido in particular.

◆ Pine nuts – these delicately flavored nuts were said to be used by the ancient Romans to boost their libido and improve their sexual staying power before indulging in their famous orgies.

◆ Truffles – these fungal balls release a pheromone, or chemical scent, that is similar to that of sexually active male pigs. Many famous lovers swear that truffles are the one true aphrodisiac.

◆ Korean, Chinese, Siberian, or American ginseng (*above*) – this helps the body to combat stress and

has long been hailed as an aphrodisiac.
◆ Asparagus – the fresh, young tips of
which Culpeper, the famous herbalist,
said "stirreth up bodily lust in man or
woman." But you should avoid the old,
woody shoots, which are thought to
have the opposite effect on your libido.

Eat these luscious vegetables with
your fingers, dipping the ends of the
asparagus tips in melted butter and
lemon juice and sucking them dry.
The reputed aphrodisiac effect may
come as much from the sensual
experience of licking your fingers

as from anything contained within the
vegetable itself.
◆ Bananas – shaped somewhat like a
phallus, this fruit may be peeled and
eaten to arouse sexual desire, or used
as a sex toy.
◆ Chocolate – this melts at body
temperature to produce sensual
explosions of flavor like liquid velvet
on the tongue – a wonderful sensation,
even if high in calories.
◆ Champagne – the bubbles enhance
the effects of the alcohol to boost your
libido, but do not overdo it.

MANY DIFFERENT FOODS
and flavors are said to be aphrodisiacs.
Try as wide a variety as you can to discover
which ones turn you and your partner on.

AND FINALLY...

A sense of humor, as well as a sense of proportion, are as important as anything else when it comes to improving your sex life.

No matter how good you feel that your sex life is at present, there is always room to improve it. If you are dissatisfied with your sexual performance in bed, however, don't despair – think positively instead. If there is a problem, you can fix it! This book is full of information, tips, illustrations, and techniques that can help you to rethink your entire attitude toward yourself, your body, and your lovemaking.

Try not to take your sex life too seriously: it is much easier to explore new lovemaking techniques and to learn to enjoy your sex life if you approach the whole subject with a good sense of humor. A willing and enthusiastic partner will add to the fun and increase your chance of mutual enjoyment.

You are likely to be disappointed if you expect miracles immediately. Like every other skill in life that is worth learning, becoming a better lover and enjoying better sex takes time, patience, and practice.

EXPRESSING YOUR NEEDS

If you are not used to talking about your needs, desires, or other similar intimate subjects, you will probably feel awkward, self-conscious, or embarrassed when first suggesting new sexual activities to your partner. If you find it difficult to talk about sex and what you want, see if you can communicate these things non-

verbally – perhaps by gently putting your partner in the position you want to try as part of your foreplay and seeing how she responds.

Alternatively, leave this book open where she will see it – or be up front about it and show her the information you are interested in and ask her what she thinks about it. You will probably be pleasantly surprised by her reactions – women want to enjoy a better, more fulfilling sex life as much as men do.

VARIETY IS THE SPICE OF LIFE

Above all, think about your sex life in the long term. Try to vary your lovemaking by regularly introducing new positions and exploring new fantasies. There are many good manuals on sexual positions that you can refer to once you have perfected the exercises and techniques in this book.

The challenge of constantly experiencing new sensations will add zest to your love life and give your relationship a welcome boost. And if, at the same time, it motivates you into improving other aspects of your lifestyle – such as health, nutrition, and general fitness – you will benefit in countless ways.

Enjoying better sex is one of the best ways of diffusing stress, lifting your mood, and improving your quality of sleep. It is also a satisfying way of enhancing a long-term, loving relationship.

INDEX

Main mentions are in **bold** type.

ACKNOWLEDGMENTS

PICTURE CREDITS

All of the photographs in the book were taken by Laura Wickenden except for the following:
p.33 D. Phillips/Science Photo Library; Dr. Tony Brian/Science Photo Library
Cover photography: Laura Wickenden

Illustrators:
Mike Saunders pp.14, 15, 16, 17, 18, 19, 20, 24, 25, 62
Jenny Holmes pp.56, 58, 59
Debbi Hinks pp.60, 63, 64, 65, 66, 67, 68, 69

The publishers also wish to thank:
Models: Charlie, Kimberly, Patrick Jones, Misha, Jonathon Richards